THIS BOOK BELONGS TO :

_____

_____

WANT MORE OF THESE BOOKS ?

Check out my Author Central Page
"Ash Creations " There You will find
More Fun and Enjoyable Books to
Play with

Pleas leave a comment or review.
We always appreciate our customers
Feedback !

Questions and Customer Service
Email us at felix@wheresit.cf

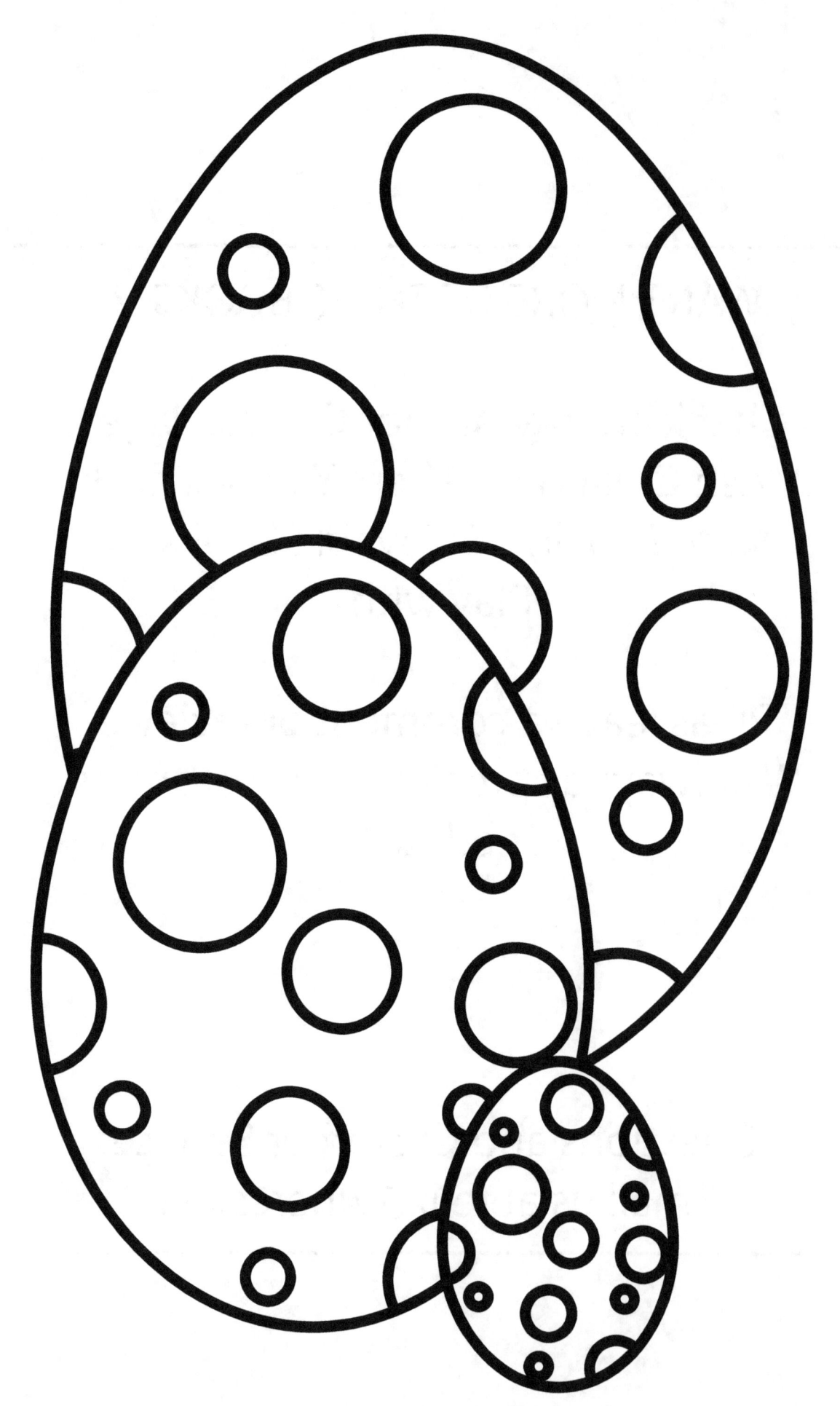

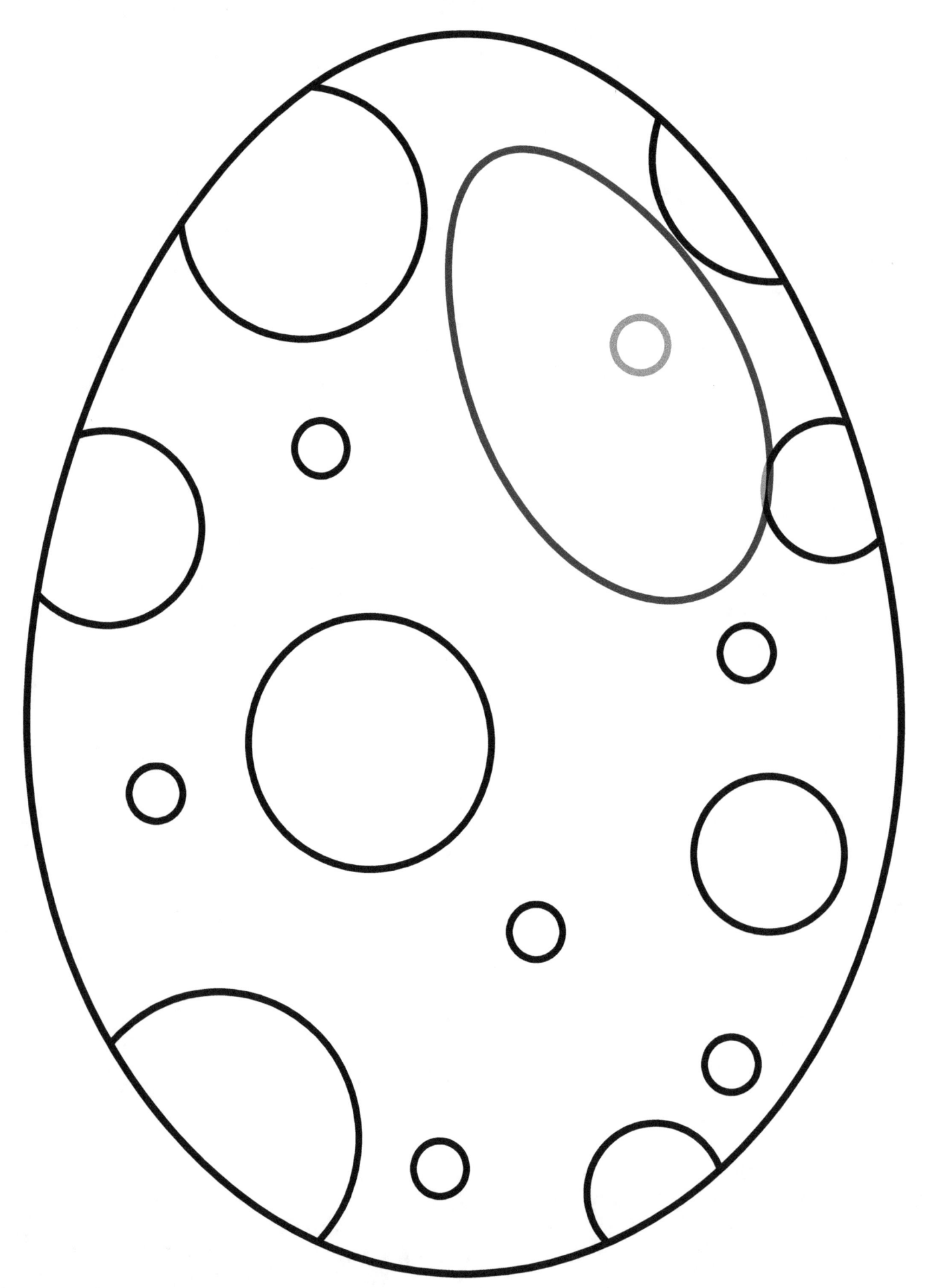

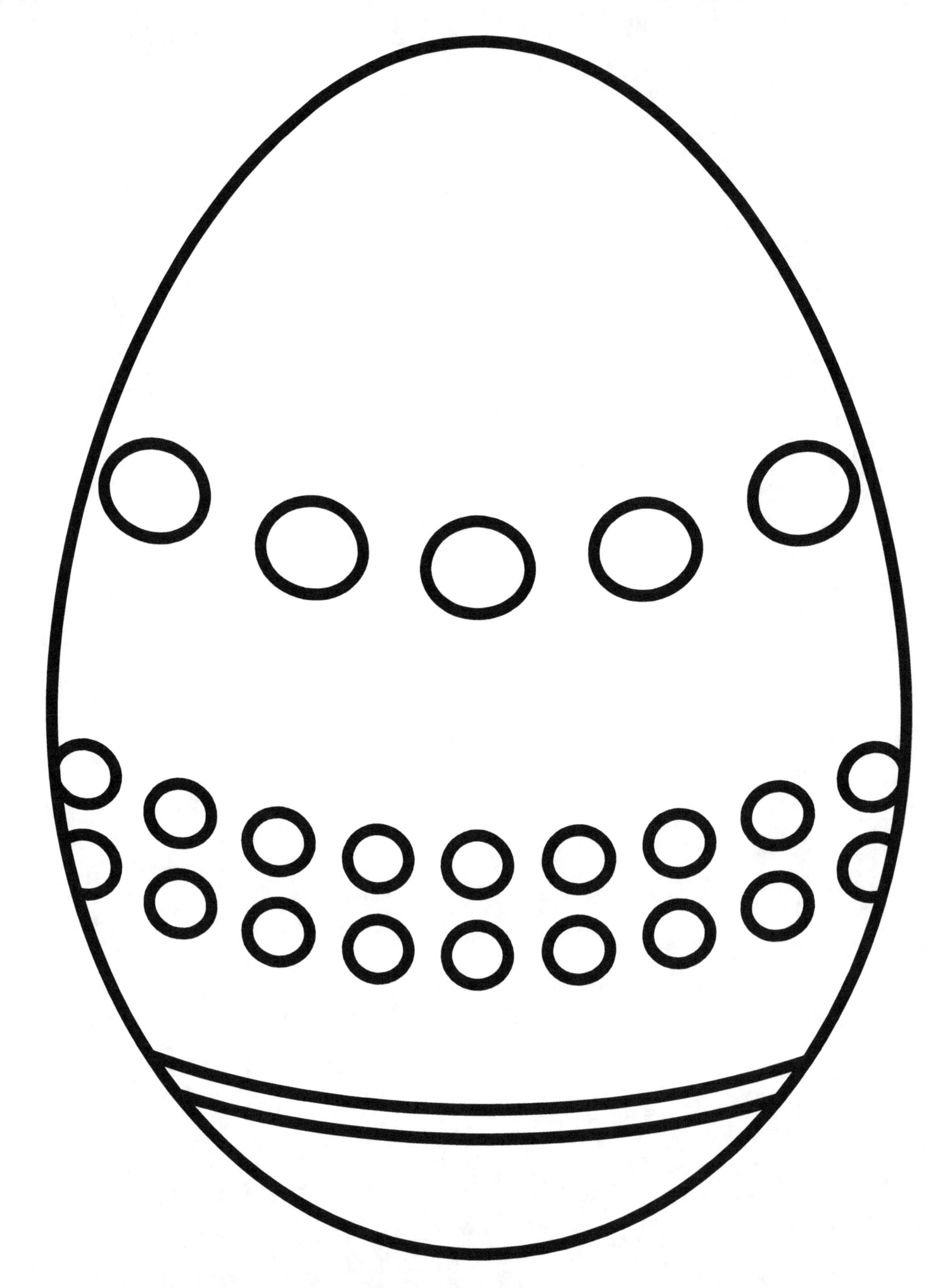

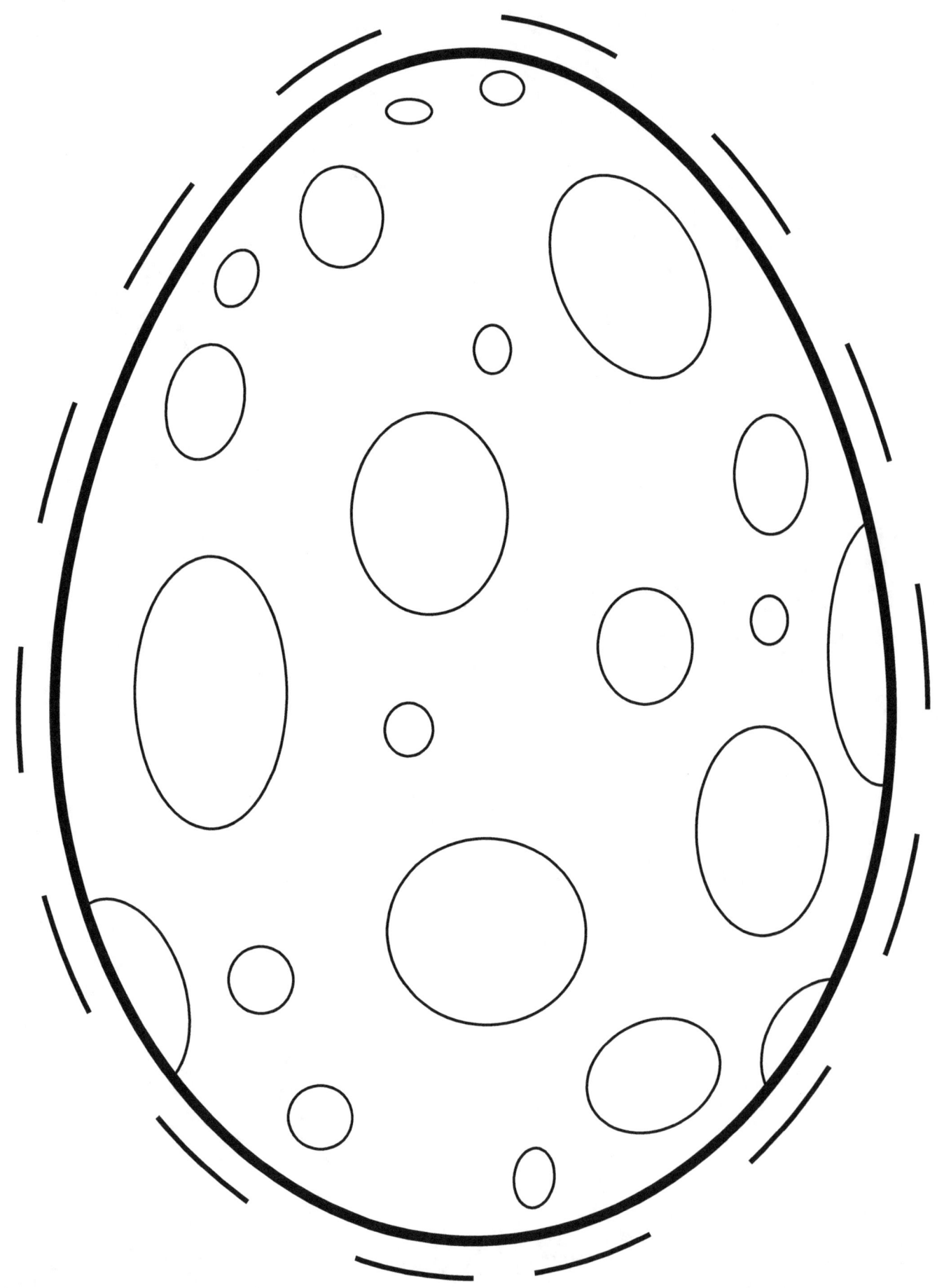